Kevin Mader

Assessment of nucleotide excision repair protein binding forces by atomic force microscopy and optical trapping

GRIN Verlag

Bibliografische Information der Deutschen Nationalbibliothek:

Die Deutsche Bibliothek verzeichnet diese Publikation in der Deutschen Nationalbibliografie; detaillierte bibliografische Daten sind im Internet über http://dnb.d-nb.de/ abrufbar.

Imprint:

Druck und Bindung: Books on Demand GmbH, Norderstedt Germany
ISBN: 978-3-638-81389-1

This book at GRIN:

http://www.grin.com/en/e-book/75136/assessment-of-nucleotide-excision-repair-protein-binding-forces-by-atomic

Assessment of Nucleotide Excision Repair Protein Binding Forces by Atomic Force Microscopy and Optical Trapping

Kevin Mader

May 11, 2007

Contents

1 Introduction

DNA is under constant repair from the damage being done from sources such as UV radiation, mutagenic chemicals, and errors made by the cell's DNA replication mechanisms. The ability for a cell to identify and repair the damaged DNA is crucial for the cell to be able to successfully function and replicate. On a systemic scale the repair is essential for maintaining long term genomic stability. When these pathways fail the usual response is for the cell to die but in some instances the damage is done in a region that causes the cell to become carcinogenic. The DNA repair enzymes are responsible for finding and correcting these mistakes.

There are many different types of damage that can be done to DNA ranging from dimerization to depurination. Each of these types of damage requires a slightly different repair mechanism. The specific type of damage that is being investigated in this proposal is pyridine dimerization which usually occurs as the result of exposure to UV radiation. The repair pathway being nucleotide excision repair which involves either the replacement or removal of a region surrounding the damaged DNA. Problems in this pathway are important in pathological conditions such as xeroderma pigmentosum which causes the skin to be over sensitive to sun exposure and a high incidence of cancer.

Also genetic engineering utilizes deletion and insertion of DNA bases into various different cells. Understanding the pathways utilized to identify the structural changes that signify damage could be utilized to construct more sensitive repair proteins. Understanding the mechanisms of repair proteins to replace the damaged DNA with the correct segment could be utilized to develop faster more efficient ways for modifying bacteria and cells in beneficial ways.

Finally understanding the mechanisms of DNA damage and repair are useful from an evolutionary standpoint. For cells and organisms to be capable of genetic adaptations to environmental forces and consequently long term survival the repair mechanisms need to work well enough to keep the genome stable, but make mistakes often enough to allow for enough diversity for survival. The balance struck between these two goals is highly variant between species. A deeper understanding of the recognition and repair of damaged DNA could provide insights into the driving factors behind the evolutionary process

2 Specific Aims

The goal of this research program is to investigate DNA repair proteins, to investigate binding force of the DNA repair protein to the defective DNA molecules using the tools developed for determining the DNA-protein complex binding strength specifically atomic force microscopy and optical trapping developed in [1, 2, 3, 4] and the magnetic tweezers methods developed in [5, 6, 7]. Studies have been done to determine the stochiometry and pathways of DNA repair [8, 9, 10]

Specifically the repair enzyme I wish to investigate is the UvrABC complex a highly conserved repair pathway for UV damage done to DNA. This complex is ideal because a significant amount of research has already been done using various tools to determine structurally how the molecule appears to bind to and recognize DNA [10, 11, 12, 13]. The goal of my research will be to determine and how strongly the complex binds to the DNA, the time-scale for binding, and how the strength of the DNA molecule is affected while the enzyme is acting. The research should provide new insights into the forces involved in recognition and replacement in the repair pathway. should confirm and strengthen results obtained from other methods of investigation while offering a significant amount of new information about the forces involved.

2.1 DNA-Protein Binding

The first set of experiments will explore the protein to DNA binding force by use of Single Molecule Force Spectroscopy (SMFS). The UvrABC complex consists of three separate proteins: UvrA, UvrB, UvrC that interact together to perform the functions of excision and repair. With this method the entire complex will not be investigated, just the UvrA and the UvrB which are responsible for damage recognition. The UV exposed dsDNA strand will be attached to the tip of the AFM and the $UvrA_2B_2$ heterotetramer protein will be linked to the surface of the stage. The stage will be slowly moved and the deflection of the AFM tip, representing the force that is being exerted, will be recorded. A better understanding of the binding forces should further elucidate the method for recognition and regions and mechanisms involved. The data would also demonstrate the sensitivity the UvrB protein has to the amount of damage done. The factors that will be modulated in the experiment are the rate of stage movement, the dsDNA segment used (both sequence and length), and the amount of UV damage done, and the linking polymers used to bind the DNA and protein to the instrument surfaces. These should provide information about the behavior of the protein in a variety of situations and isolate the DNA-protein binding from the other factors such as linker elasticity and electrostatic interactions between DNA segments and the protein.

2.2 DNA Strength during UvrABC Binding

The second set of experiments will measure the tensile properties of the DNA while the complex is repairing damage. The dsDNA molecule will be attached to the stage and the AFM tip. The molecule will be stretched until taut by moving the stage away from the tip. As the complex is added the tension will be measured by recording the deflection of the tip corresponding to damage recognition, unwinding, and excision. The experiment will be to perform this process in the presence of low concentrations of UvrABC to record the changes in structural properties during nucleotide excision repair. Given that neither the UvrD nor the DNA polymerase I are present the repair process will not be able to complete and the DNA strand will be left in a structurally weak state. This can be further investigate by simply performing the stretching properties of the DNA molecule before and after the complex is added. Since it is known that the rate at which the repair occurs is ATP dependent [11] the concentration of ATP will be modulated to verify the observed events have a concentration dependence. Additionally the concentration of the UvrABC complex will be kept in very low concentrations ($\approx$nM) in order to assure that only one complex is interacting at a given time, single molecule sensitivity.

2.3 Optical Tweezers

The third specific aim is to repeat the measurements of the first two experiments using a different tool in order to verify the numbers obtained. All of the experiments done on the AFM setup will also be conducted using an optical tweezers setup. This will allow for the results to be corroborated and the errors from setup-specific sources to be minimized. Optical trapping methods will be used to bind the DNA to a polystyrene bead that can be held in a fixed position using a laser. The deflections of this bead can be used to measure the forces being exerted on the bead. Furthermore magnetic tweezers offer an even higher force resolution if in the event the forces observed are too small to be seen by either AFM or OT. The setup for a magnetic tweezer is similar to optical tweezers, but the bead is a magnetic material and a magnetic field is used to hold the bead in place. The distance the bead moves away from resting is related to the force being exerted. Finally magnetic tweezers offer the ability to measure twisting and and exert torque on the DNA strand, which we do not initially plan on doing, but could provide an interesting extension to the research.

3 Background and Significance

The failure of the mechanisms of DNA repair play a crucial role in many diseases including cancers. If these mechanisms were better understood they could be utilized to develop stronger treatments. There are 3 primary types of damage that is done to the DNA: depurination, deamination, and dimerization. Each of these types of damage occurs in a different way and requires different repair mechanisms to fix. This proposal only deals with UV induced dimerization.

3.1 Dimerization

When UV light or some chemicals interact with DNA, they can interact with a pyrimidine base (cytosine and thymine) in a photocycloaddition reaction to form a cyclobutane-type pyrimidine dimer as shown in Figure 1. The formation of this dimer means that the adjacent thymine bases are no longer bound to the adenine bases on the complementary DNA molecule but to themselves. This disrupts the double helix nature of DNA as shown in Figure 2. After this occurs a protein complex such as a DNA polymerase trying to unzip the DNA molecule would not be able to get past this spot and the functions of the cell dependent on this sequence would be unable to proceed [14].

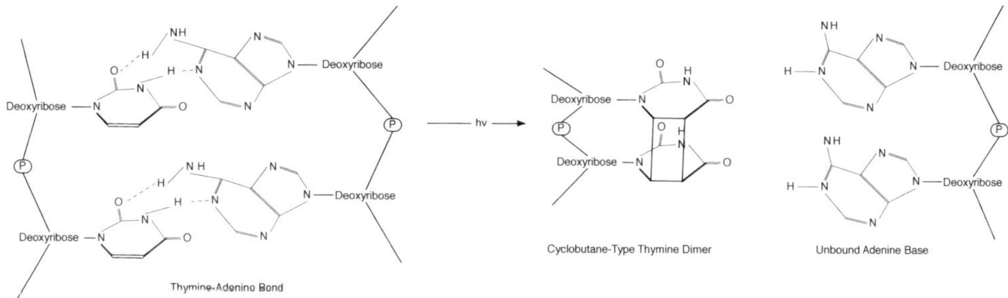

Figure 1: Cyclobutane-type Pyrimidine Dimer Formation between adjacent Thymine bases

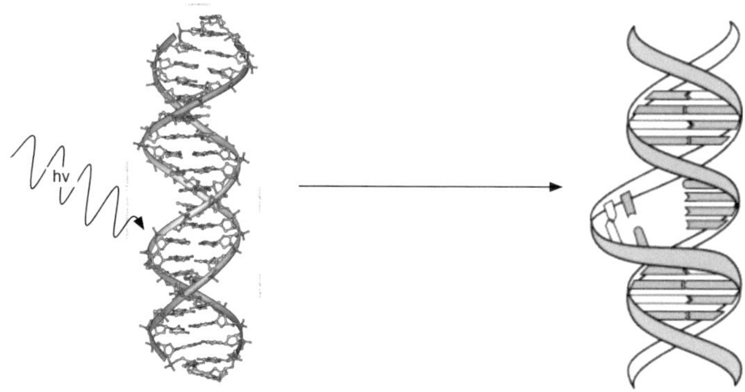

Figure 2: Change in DNA structure due to dimerization

3.2 Nucleotide Excision Repair Pathway

The nucleotide excision repair pathway is shared by a wide variety of species ranging from bacteria to mammals. The process is performed by several complexes known collectively as UvrABC. The three proteins involved UvrA, UvrB, and UvrC act co-operatively to track, locate the damaged region, and perform the excision on the DNA molecule. The tracking is done using a $UvrA_2B_2$ heterotetramer driven by UvrB to locate the sites of a structural deformation in the DNA such as one caused by dimerization shown in Figure 2. The interaction the UvrB has with the lesion causes the UvrA-B complex to change shape and unwind a portion of the DNA adjacent to the lesion. The unwound DNA wraps around UvrB-DNA complex which through a currently debated mechanism recruits UvrC which cleaves damaged section at the 4th or 5th phosphodiester bond $3'$ to the lesion and cleaving the 8th phosphodiester bond $5'$ to the site. The UvrBC-DNA complex formed is stable until the UvrD (a DNA helicase) binds and displaces the damage containing section of the strand. The gap in the strand is then filled by DNA polymerase I. The final steps of repair are completed by DNA ligase [10, 15, 13].

3.3 Atomic Force Spectroscopy

The general setup for atomic force microscopy involves the use of a flexible cantilever with a tip and mirror attached at the end. The tip is made out of a material that is predictively interactive with the sample being probed and the mirror acts to amplify the tiny motion of the cantilever tip to an easily measurable quantity because small deviations in position and orientation of the mirror correspond to a small angle change of the beam path of the laser. This small change in angle is scaled into a change in beam position by the distance away the detector is from the sample. The force is determined by back calculating the position change on the detector to a angle change on the mirror to a small position change in the cantilever. This position change can be translated into a force because the cantilever can be modeled as a spring where the amount of deformation is linearly proportional to the applied force.

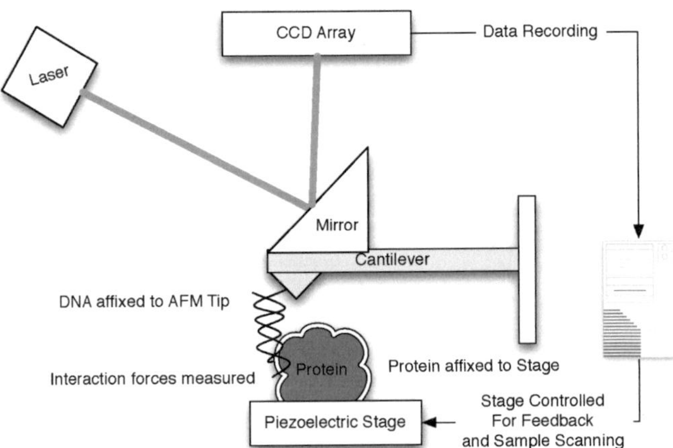

Figure 3: The slightly modified AFM setup for use in performing force spectroscopy measurements on DNA-protein complexes

3.4 Optical Trapping

The setup for optical trapping uses a polystyrene bead held in place by optical trapping forces. The optical trapping forces act to push the bead toward the focus of the laser. The distance the bead is away from the focus can be converted into the force that must be acting on the bead by use of simple models.

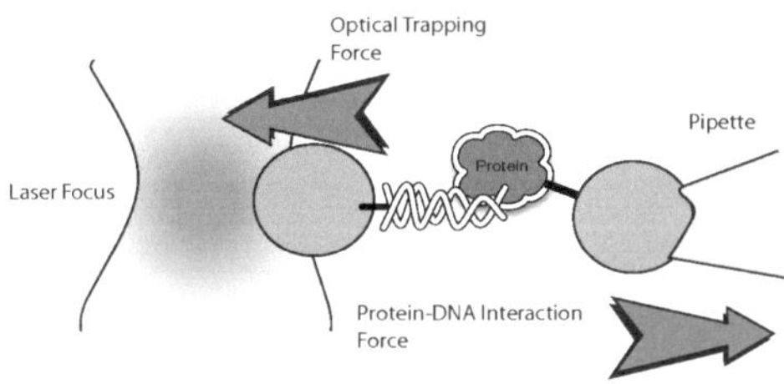

Figure 4: The Optical Trapping Setup Being Used for this Experiment

4 Preliminary Studies

The use of AFM and optical tweezers for force spectroscopy provide piconewton force, nanometer distance, and millisecond time resolution. The selection or use of either method will be based on the approximate force of the interaction and the likely range of forces to be probed although for most of the experiments both setups will be used to collaborate the results obtained. The two methods provide a slightly different range and accuracy of measurements. Both models can be approximated using a spring model. The spring constant for the cantilever arm used in a similar DNA-protein interaction was $\approx 14\frac{pN}{nm}$. [1]. This stiffness in combination with a sub-nanometer sensitivity and the ability to see displacements of up to 1mm results in the ability to see forces from 10pN-10μN [3]. The Optical Tweezers setup has a much lower effective spring constant of $< 1\frac{pN}{nm}$. This allows for a much higher sensitivity up to the subpiconewton range, but in combination with the beam-shape of the laser limits the maximum force accurately measurable is $\approx 200pN$. The limit placed on both setups for maximum force is the strength of the linking molecule to the tip or DNA strand. The unbinding force for this is $\approx$550pN so forces close to or above this number cannot be easily probed. A nice diagram illustrating the differences using a spring model is shown in Figure 5.

Studies using other proteins bound to DNA have shown an interaction force ranging between 40-200pN [16]. The interaction force represents a mean of the results so values above the 200pN need to be probed to obtain the result. With this information the AFM becomes the more useful tool since the forces being probed could easily exceed the 200pN maximum for the optical tweezer instrument.

The standard sorts of forces that would be expected in a DNA-protein binding are electrostatic forces, dipole-dipole forces, and the hydrogen bonding forces. Each of these forces represent different portions of the DNA and protein interacting and have different mechanical characteristics and strengths associated with it. Additionally the type of force is capable of elucidating information on the DNA sequence specificity of the binding.

4.1 Force Spectroscopy

Force spectroscopy involves recording the force being exerted on the instrument against the displacement of the stage. The effective result is you can obtain a sort of elasticity factor on the complex being probed.

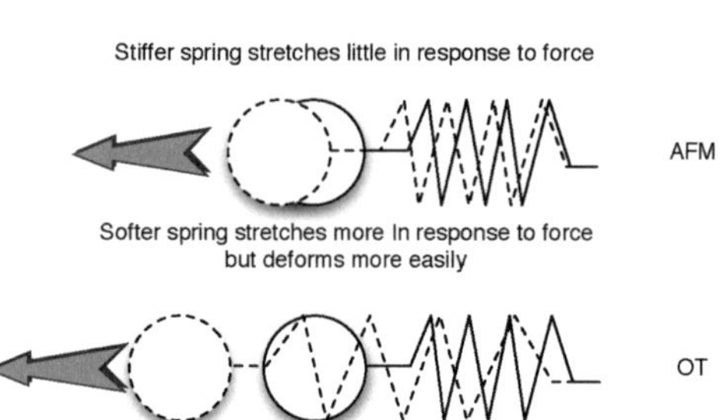

Figure 5: Comparison of Spring Model for AFM and OT

Electrostatic Interactions Positively charged amphiphilic proteins tend to interact with the negatively charged phosphate groups on the DNA. Although this interaction is nonspecific there are several transcription factor.

4.2 Dynamic Force Spectroscopy

Dynamic force spectroscopy operates moving the stage at a velocity and recording the force. This is in contrast to force spectroscopy which records the force at specific displacements. DFS is a better suited method for investigating many of the properties of the interaction since it is better aligned with the thermodynamic properties of the system. WIth a slowly varying force rate the thermal fluctuations in the system have a greater chance of spontaneously overcome the energy barrier for the bond. The DFS method allows for measurement of the lifetime of the complex, the thermal off rate, and can be used to investigate energy barriers in the transitional states of the complex [2].

4.3 DNA Stretching

Another type of experiment that has been performed on many similar substrates is the stretching of the DNA molecule between the instrument and the stage. The force versus distance response of the stretching is recorded and is fairly well understood. The DNA begins to change shape from the regular dsDNA to S-DNA which is 1.7x as long as the original strand. The winding of the DNA also begins to change as this transition is made which can cause some proteins that are sensitive to larger structure to unbind [5].

4.4 Groove Binding

Some of the DNA-protein interactions involve the peptide binding in grooves formed in the double stranded helical structure of the DNA. The two classes of groove binding are minor and major. Minor groove binding involves the interaction of several small charged particles in the groove of the double helix. This binding does little to change the conformation of the DNA molecule and operates on a weaker bond. The major groove binding consists of strong electrostatic interactions between the peptide and the backbone of the DNA that cause a larger conformational change in the DNA [3]. The groove binding mechanisms are also sensitive to the stretching and winding of the DNA described in the last section.

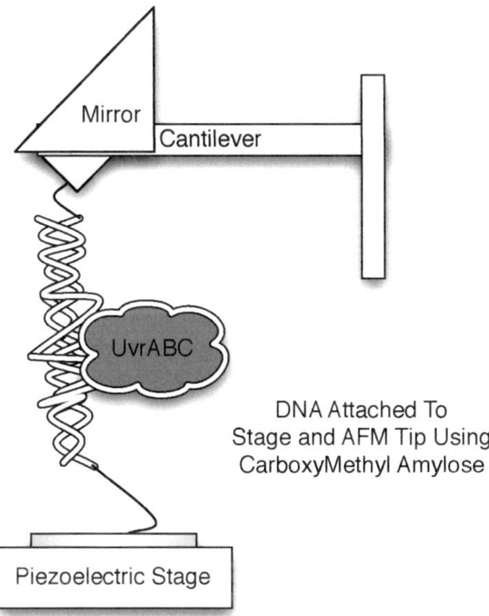

Figure 6: The AFM Setup for force spectroscopy measurements on the DNA strand while the protein is interacting with the damaged region

5 Research Design and Methods

5.1 Sample Preparation

The samples of DNA will be prepared in a number of different ways in order to determine the affinity for the protein complex for the DNA. The only real requirement on the strand are that there must be parts of the sequence that have consecutive pyrimidine bases so the dimer formation can occur. The next requirement is the DNA must be at least 8 base-pairs long because the excision pathway removes a total of 14-15 bp (cleavage of the 4th or 5th that are 3′ and the 8th that is 5′, to the damage site). For simplicity of measurement the first several strands will be a synthetically created thymine rich strand. The synthetic strands will be generated according to the methods shown in [9], purchasing from Trilink Biotechnogies (San Diego, CA) with the minor change being that for the final experiments none of the fluorescent dyes will not be attached since they would likely alter the structural properties. However, for calibration purposes labeled DNA sequences would be useful to verify attachment of the DNA to the tip or bead and possibly to verify the success of other similar setup related tasks. Some of the other strands that will be used will be actual segments of DNA from *E. Coli*. The subsequent strands will be determined based on the success of the thymine rich strands and the contro

5.1.1 DNA Damage

The DNA will be irradiated with 266 nm pulses generated by the fourth harmonic of a Nd:YAG laser. The pulses will be sent at a 10 Hz repetition rate with an energy density of approximately 200 nJ / cm^2. The forth harmonic will be obtained using a quasi-phase matched cascaded system with two periodically poled Lithium Niobate ($LiNbO_3$) nonlinear crystals. The use of this source will require optics made out of BaF_2 since standard glass optics absorb too strongly in this regime. The wavelength of the excitation source was chosen because the ideal excitation wavelength to cause dimerization damage is located at 260nm, but the range between 160nm-300nm still functions efficiently enough to be useful for the experiment [17]

5.1.2 DNA Damage Quantification

Although the duration of exposure will be measured, this is not a quantifiable enough number to know how much damage has been done. An IR spectra can be taken of the DNA sample in order to determine the amount of dimerization has occurred. Several spectra will be taken on unexposed samples in order to determine a baseline spectra. Then each of the irradiated samples will have the spectra taken and these will be compared to the baseline samples and the differences will be quantified using FTIR spectroscopy (Perkin-Elmer 1720X) with specific interest in the change in the peaks at 1632 cm^{-1}, 1664 cm^{-1},1693 cm^{-1}. These peaks shrink as the number of dimers formed in the sample increase [18]. This will be done for each different sequence and sample length of DNA because the spectra changes as the composition and length of the strand change. The number of dimers formed can be roughly estimate from the spectral change and will be attempted to be held constant.

5.1.3 Binding DNA to AFM Tip

The cantilevers being used will be a Si_3N_4 made by Veeco Microscopes in Mannheim, Germany. They offer several different setups with different stiffness properties. The first setup used will be the Model A MLCT-AVHW since that was successfully used in similar experiments. The first step is the tip will be cleaned by using a mercury UV lamp for 30 minutes followed by washing the cleaned tip with ethanol and 90°C deionized water. The first linker being used is CMA (carboxymethylamylose). This will be attached to the surface by incubating the tip for 10 minutes with a solution made up of 10 mg of carboxymethylamylose mg (Sigma) + 4 mg NHS (N-hydroxysuccinimide, Aldrich) + 17 mg of EDC (1-ethyl-3-[3-(dimethylamino)propyl]carbamide, Sigma) dissolved in 200 mL of PBS (Phosphate Buffered Saline) [16].

In order to attach the DNA molecule to the prepared AFM tip a small volume ($\approx 20\mu$L) solution containing the DNA and PBS should be incubated with the tip for a period of around 90 minutes. The time for this binding to occur varies and for the first experiment fluorescently labeled DNA will be used to verify the attached was successful.

5.1.4 Binding DNA to Polystyrene Bead

In order for the DNA strand to be bound to the polystyrene bead several things must be done. The first step is bind biotin to a terminal end of the strand. The strand is then placed in vicinity of a streptavidin-coated 2.67 μm polystyrene bead. The bead is held in place by a suction pipette

5.1.5 Protein Sample Preparation

The protein samples will be obtained from genetically modified His-tagged *B. subtilis* UvrA and UvrB gene as described in [11]. Similar methods can be used to obtain the UvrC protein. Once all of the proteins are obtained and purified. Additionally using this method allows for the possibility of adding point mutations to the protein sequence. These point mutations could be used to verify the structure that causes the forces by disrupting portions of it.

5.1.6 Binding Protein to Slide

There are two techniques that will be used to bind the protein to the movement of the stage. Both methods involve the attachment of the linker molecule to external structures of the protein rather than a synthetic attachment to one of the terminal ends of the polypeptide. This was done intentionally because some research has shown that the C-terminal end zinc-finger of the UvrA protein plays a critical role in the identification of the damage region [19].

The first method that will be used is the similar to the method used to attach the DNA to the AFM tip. The method requires the use of Aminoslides from Quantifoil Micro tools GmbH (Jena, Germany). The CMA linker is attached by incubating the Aminoslide for 10 minutes with a solution made up of 10 mg of carboxymethylamylose mg (Sigma) + 4 mg NHS (N-hydroxysuccinimide, Aldrich) + 17 mg of EDC (1-ethyl-3-[3-(dimethylamino)propyl]carbamide, Sigma) dissolved in 200 mL of PBS. The UvrB protein can be attached to the CMA by further incubating the prepared slide with a small volume ($\approx 10\mu$L) of the protein diluted to about 30mM in PBS solution for around 90 minutes [16].

The second technique for attaching the protein to the stage is to evaporate the protein on a gold surface. The protein samples are prepared and attached to the linking molecule 1,8-diamino-3,6-dioxaoctane. The solution is then incubated for 24 hours at 4°C on the gold surface which has been freshly evaporated onto the stage [1].

5.2 AFM Setup

The final AFM setup shown in Figure 7 shows how the different elements in the system are interacting when CMA is being used as the linking agents. The other compounds have a similar functionality. The AFM instrument to be used will be the Model A MLCT-AVHW by Veeco Microscopes (Mannheim, Germany). The cantilever measurements will be calibrated using the thermal fluctuation method [20].

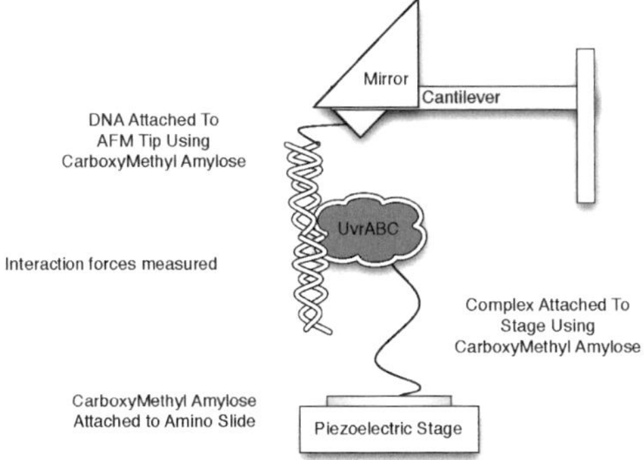

Figure 7: The AFM Setup for force spectroscopy measurements on the DNA-protein complexes

5.3 Optical Trapping Setup

The optical trapping setup will be a standard inverted microscope setup with a high numerical aperture ($\approx$1.2) in order to focus the beam tightly. The same Nd:YAG laser that has a fourth harmonic used for the DNA damage can be used as the trapping laser in this experiment. A standard 1340 $\times$ 100 pixel CCD camera (Roper Scientific, Trenton, New Jersey) will be calibrated and then used in combination with a halogen light source in order to image and track the position of the bead. The beam splitter from Kaiser Optical with a notch filter at the 1064nm wavelength to prevent the laser light from saturating or possibly damaging the CCD. Alternatively a simpler setup could be constructed where instead of imaging the bead movement you could image the beam at the output and calculate the displacement of the bead by how much the beam has been shifted.

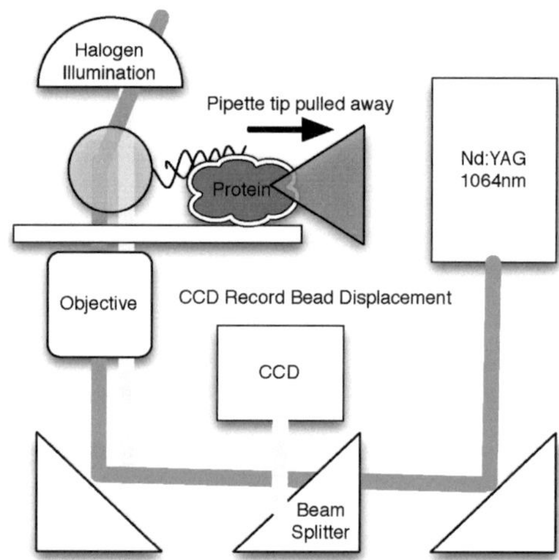

Figure 8: An example of the optical trapping setup where the red line represents the laser line and the yellow line represents the wide-field light used for viewing the sample and measuring the displacement of the bead.

5.4 Expected Sources of Error

Most of the errors that are likely to be encountered are due to the experiment setup. In figure 9 the errors that likely enter through uncertainty of mechanical properties are shown. These errors will be minimized by utilizing as many different setup configurations as possible for the linker lengths and materials, AFM tips, and the use of optical trapping. The errors will be further understood by switching through several different DNA molecule sequences and length.

Some of the other problems we are likely to encounter is that the environment setup in this experiment although chemically similar to, is significantly different from the nuclear environment the UvrABC complex usually functions within. The absence of the usual enzymes, coenzymes, proteins, and other structures that may interact with the DNA molecule that would be present in situ could largely affect the kinetics and mechanics observed in the experiment when conducted in vitro.

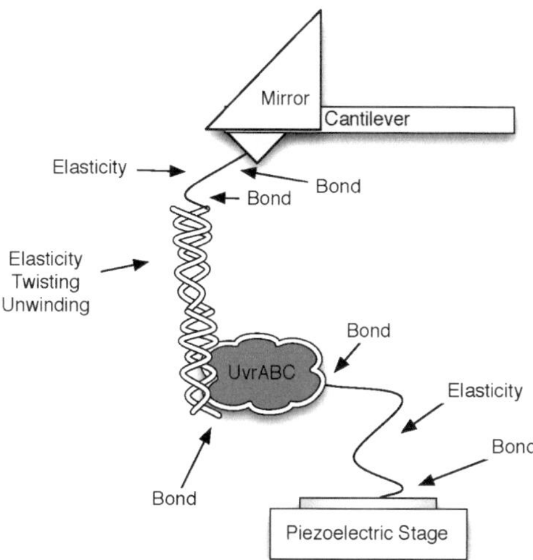

Figure 9: The regions of the setup likely to have an impact on the force measurements

5.5 Experimental Analysis

The thermodynamic nature of the experiments means that rather than trying to analyze single absolute values, the probability distribution of the experiments will be analyzed. The histogram of results produced will be fit to the expected probability distribution for the event being measured.

5.5.1 DNA-Protein Rupture Forces

A first order approximation for the rupture forces is a one-barrier binding potential [16]. The

$$p(F) = k^*_{off} \exp\left(\frac{F\Delta x}{k_B T}\right) \frac{1}{\dot{F}} \exp\left(-k^*_{off} \int_0^F \exp\left(\frac{F\Delta x}{k_B T}\right) \frac{1}{\dot{f}} df\right) \quad (1)$$

The equation representing the probability of an unbinding event for a given applied force and loading rate. The known parameters such as k_B is Boltzmann constant, T is the absolute temperature. Δx, F, $\dot{F}$ are all obtained from the experimental data. This leaves only k^*_{off} representing the thermal off-rate which must be fit to the data. Similar methods can be used to fit the complex lifetime (τ) parameter to the results.

5.5.2 Force Spectroscopy

The second sort of analysis that can be done is looking at the force recorded by the instrument while retracting the stage. This gives information on the different stages or types of binding involved since usually there are multiple transitions that occur during this lengthening process. These transitions could represent changes in the DNA structure such as melting where the double-stranded structure begins to dissociate, or actual changes in the bonding of the complex to the DNA. The presence of major instead of minor groove binding would have different effects on the structural properties of the DNA strand.

Literature Cited

[1] Eckel, R., Ros, R., Ros, A., Wilking, S. D., Sewald, N., and Anselmetti, D. Identification of Binding Mechanisms in Single Molecule-DNA Complexes. *Biophys. J.* **85**(3), 1968–1973 (2003).

[2] Norbert Sewald, Sven D. Wilking, R. R. D. A. Probing dna-peptide interaction forces at the single-molecule level. *Journal of Peptide Science* **12**(20), 836–842 (2006).

[3] Ros, R., Eckel, R., Bartels, F., Sischka, A., Baumgarth, B., Wilking, S. D., Puhler, A., Sewald, N., Becker, A., and Anselmetti, D. Single molecule force spectroscopy on ligand-dna complexes: from molecular binding mechanisms to biosensor applications. *Journal of Biotechnology* **112**(1-2), 5–12 (2004).

[4] Rainer Eckel, Sven David Wilking, A. B. N. S. R. R. D. A. Single-molecule experiments in synthetic biology: An approach to the affinity ranking of dna-binding peptides. *Angewandte Chemie International Edition* **44**(25), 3921–3924 (2005).

[5] Gore, J., Bryant, Z., Nöllmann, M., Le, M. U., Cozzarelli, N. R., and Bustamante, C. Dna overwinds when stretched. *Nature* **442**(7104), 836–839 (2006).

[6] Strick, T. R., Allemand, J., Bensimon, D., Bensimon, A., and Croquette, V. The Elasticity of a Single Supercoiled DNA Molecule. *Science* **271**(5257), 1835–1837 (1996).

[7] Smith, S., Finzi, L., and Bustamante, C. Direct mechanical measurements of the elasticity of single DNA molecules by using magnetic beads. *Science* **258**(5085), 1122–1126 (1992).

[8] Lindahl, T. and Wood, R. D. Quality Control by DNA Repair. *Science* **286**(5446), 1897–1905 (1999).

[9] Fore, S., Laurence, T. A., Yeh, Y., Balhorn, R., Hollars, C. W., Cosman, M., and Huser, T. Distribution analysis of the photon correlation spectroscopy of discrete numbers of dye molecules conjugated to dna. *Selected Topics in Quantum Electronics, IEEE Journal of* **11**(4), 873–880 (2005).

[10] Waters, T. R., Eryilmaz, J., Geddes, S., and Barrett, T. E. Damage detection by the uvrabc pathway: Crystal structure of uvrb bound to fluorescein-adducted dna. *FEBS Letters* **580**(27), 6423–6427 (2006).

[11] Eryilmaz, J., Ceschini, S., Ryan, J., Geddes, S., Waters, T. R., and Barrett, T. E. Structural insights into the cryptic dna-dependent atpase activity of uvrb. *Journal of Molecular Biology* **357**(1), 62–72 (2006).

[12] Yamagata, A., Masui, R., Kato, R., Nakagawa, N., Ozaki, H., Sawai, H., Kuramitsu, S., and Fukuyama, K. Interaction of UvrA and UvrB Proteins with a Fluorescent Single-stranded DNA. Implication for slow conformational change upon interaction of UvrB with DNA. *J. Biol. Chem.* **275**(18), 13235–13242 (2000).

[13] Nakagawa, N., Sugahara, M., Masui, R., Kato, R., Fukuyama, K., and Kuramitsu, S. Crystal Structure of Thermus thermophilus HB8 UvrB Protein, a Key Enzyme of Nucleotide Excision Repair. *J Biochem (Tokyo)* **126**(6), 986–990 (1999).

[14] Lyamichev, V. I., Frank-Kamenetskii, M. D., and Soyfer, V. N. Protection against uv-induced pyrimidine dimerization in dna by triplex formation. *Nature* **344**(6266), 568–570 (1990).

[15] Esther E.A. Verhoeven, Claire Wyman, G. F. M. J. H. H. and Goosen, N. Architecture of nucleotide excision repair complexes: Dna is wrapped by uvrb before and after damage recognition. *The EMBO Journal* **20**, 601–611, February (2001).

[16] Kuhner, F., Costa, L. T., Bisch, P. M., Thalhammer, S., Heckl, W. M., and Gaub, H. E. Lexa-dna bond strength by single molecule force spectroscopy. *Biophysical Journal* **87**(4), 2683–2690 (2004).

[17] Matsunaga T, Hieda K, N. O. Wavelength dependent formation of thymine dimers and (6-4) photoproducts in dna by monochromatic ultraviolet light ranging from 150 to 365 nm. *Photochem. Photobiol.* **54**(3), 403–410, September (1991).

[18] Schreier, W. J., Schrader, T. E., Koller, F. O., Gilch, P., Crespo-Hernandez, C. E., Swaminathan, V. N., Carell, T., Zinth, W., and Kohler, B. Thymine dimerization in dna is an ultrafast photoreaction. *Science* **315**(5812), 625–629 (2007).

[19] Croteau, D. L., DellaVecchia, M. J., Wang, H., Bienstock, R. J., Melton, M. A., and Van Houten, B. The c-terminal zinc finger of uvra does not bind dna directly but regulates damage-specific dna binding. *Journal of Biological Chemistry* **281**(36), 26370–26381 (2006).

[20] Hutter, J. L., B. J. Calibration of atomic-force microscope tips. *Rev. Sci. Instrum.* **7**, 1868–1873 (1993).

[21] Jurgen H.G. Huisstede, Vinod Subramaniam *, M. L. B. . Combining optical tweezers and scanning probe microscopy to study dna-protein interactions. *Microscopy Research and Technique* **70**(1), 26–33, November (2006).

[22] Vassylyev, D. G., Kashiwagi, T., Mikami, Y., Ariyoshi, M., Iwai, S., Ohtsuka, E., and Morikawa, K. Atomic model of a pyrimidine dimer excision repair enzyme complexed with a dna substrate: Structural basis for damaged dna recognition. *Cell* **83**(5), 773–782 (1995).

[23] Philip Tinnefeld, Mike Heilemann, M. S. Design of molecular photonic wires based on multistep electronic excitation transfer. *ChemPhysChem* **6**(2), 217–222, January (2005).

[24] Mike Heilemann, Robert Kasper, M. S. Dissecting and reducing the heterogeneity of excited-state energy transport in dna-based photonic wires. *Journal of the American Chemical Society* **128**(51), 16864–16875, December (2005).

[25] Krieger, F., Fierz, B., Bieri, O., Drewello, M., and Kiefhaber, T. Dynamics of unfolded polypeptide chains as model for the earliest steps in protein folding. *Journal of Molecular Biology* **332**(1), 265–274 (2003).

[26] Hannes Neuweiler, Marc Löllmann, M. S. Dynamics of unfolded polypeptide chains in crowded environment studied by fluorescence correlation spectroscopy. *Journal of Molecular Biology* **365**(3), 856–869, January (2007).

[27] Buscaglia, M., Lapidus, L. J., Eaton, W. A., and Hofrichter, J. Effects of Denaturants on the Dynamics of Loop Formation in Polypeptides. *Biophys. J.* **91**(1), 276–288 (2006).

[28] Fierz, B. and Kiefhaber, T. End-to-end vs interior loop formation kinetics in unfolded polypeptide chains. *Journal of the American Chemical Society* **129**(3), 672–679 (2007).

[29] Hagen, S. J., Carswell, C. W., and Sjolander, E. M. Rate of intrachain contact formation in an unfolded protein: temperature and denaturant effects. *Journal of Molecular Biology* **305**(5), 1161–1171 (2001).

[30] Fore, S., Yeh, Y., Huser, T., Balhorn, R., Cosman, M., and Laurence, T. Single molecule analysis of DNA synthesis using non-classical photon statistics. *APS Meeting Abstracts* **1**, 8012, March (2004).

[31] Daggett, V. and Fersht, A. The present view of the mechanism of protein folding. *Nature Reviews Molecular Cell Biology* **4**(6), 497–502 (2003).

A Alternative Methods

A.1 Scanning Probe

One possible alternative method that would allow for more detailed measurements on the DNA-protein interaction would be to use the optical tweezers and scanning probe combined method [21]. This method involves using the optical tweezers setup to hold the DNA strand taut and scanning a probe along the DNA strand. The tension along the DNA strand would be measured by the deflection of the optically trapped bead attached to the end of the strand. The could give more detailed information on the regions of the DNA molecule the complex was most sensitive too.

A.2 Mutations

Point mutations could be made the Uvr proteins at the likely sites of binding as determined by correlated the bond strength as determined by experimentation to the likely bonding sites as shown in the crystallized structures

A.3 Species of Protein

Although the UvrABC pathway is highly conserved from species to species, there are slight variations in the actual structure and function of these molecules obtaining the protein from other species could provide important distinctions between these.